THE
FIRE~
HOUSE
TRIVIA
BOOK

by

WILLIAM J. GEIS
Lieutenant F.D.N.Y.

Quinlan Press
Boston

Published by Quinlan Press
131 Beverly Street
Boston, MA 02114
(617) 227-4870
1-800-551-2500

Library of Congress
Catalog Card Number 87-81980
ISBN 1-55770-002-8

Printed in the United States of America
September, 1987

TO THE BRAVEST

About the Author

William J. Geis was appointed to the FDNY in 1968. He served in the Harlem section of the city until his promotion to lieutenant in 1981. He was then transferred to the South Bronx and at present is assigned to Ladder Company 17. He has twice been cited for conspicuous duty. He lives in Orange County with his wife, Alexis, and son, Billy.

CONTENTS

INTRODUCTION

It is my hope that those firefighters and friends of firefighters who read this book, do so merely for enjoyment. *Firehouse Trivia* was not written as a study manual. Although the material presented is accurate as far as I know, the book should not be considered a source. There are more scientific works that can be used for research.

Almost all of the questions and answers that follow can be verified. Some, however, are my own opinions and assumptions based on nineteen years of firefighting. I believe my answers in these areas are correct, and I stand by them. If, because of your own experience you disagree, that is understandable.

Firefighting is a complex field of study. I have listed only the interesting facets of this noble profession, and present them here for your pleasure.

Enjoy!

W.J.G.

History

In November of 1968, I walked into the firehouse for the first time. I was fresh out of Training School and ready for fire duty. It is not that long ago, but in a way, it was a different era.

The alarms came in on telegraph bells then; now a computer prints the address of the fire or emergency. The computer is more efficient but certainly not as glamorous. We slid the poles then; today most firefighters use the stairs. The turnout coats were either rubber or canvas; today they are Nomex.

The rig was a 1953 American La France and it was out of service more than it was in service. Today we have a modern fleet of trucks consisting mostly of Mack pumpers, Mack Tower Ladders, and Seagrave and American La France aerial ladders.

History

In those days, the air masks were kept in boxes on the apparatus and almost never used. It was considered part of the job to take a "good feed" of smoke at a fire. Now, with plastics and chemicals so widely used, it is foolish to enter a fire-building without an air tank strapped to your back.

So much has changed in the few years that I have been a firefighter. But that is how it has been throughout the history of the fire service. From ancient Rome to medieval England to Colonial America to present times, there has been a steady improvement in the quality of service that firefighters have provided.

Here are some trivia questions on the history of firefighting. You may find some easy and some difficult but I think you will agree that they are all interesting.

History

1. Who was Emperor of Rome when the first fire prevention regulations were issued?

2. What is the origin of the word "curfew"?

3. What type roof did London's first mayor outlaw since it was considered a fire hazard?

4. In China, in the 13th century, how many firemen would turn out for a fire in the Celestial City?

5. Who patrolled Nottingham England's streets on firewatch during the British Civil War?

6. What was the fire watchman in Boston during the 1650's called?

7. Both N.Y.C. and London formed paid Fire Departments in what year?

8. Who fought the great London fire of 1666?

9. What town in England organized the first paid Fire Department?

10. In what city was gunpowder first used to destroy a building in a fire's path of travel?

11. What did the first paid Fire Dept. in America consist of?

12. What was the fine in Boston for failure of a citizen to respond to a fire with his fire bucket?

13. What was a bed key used for?

14. What were the mutual fire societies of Boston the forerunners of?

15. What were the two reasons that the larger cities changed from volunteers to paid firemen?

16. Why did Boston firemen object to the introduction of suction engines?

17. What was the principle source of water in Boston in the 1800's?

18. Where and when was the first fire college founded?

19. What state organized the first state fire school?

20. When were adjustable fog nozzles first used?

21. Who developed the first municipal alarm system?

22. Who invented the telegraph fire alarm box?

23. What country had the first known fire brigade?

24. What Englishman received the first patent for a fire engine?

25. What two brothers from Holland designed a leather hosepipe in 1672?

26. Which Fire Department ordered the first gasoline motor pumper?

27. In what year did it become practical for Fire Dept.'s to install radio communications?

28. Who ordered the first fire walls to be built?

29. In what city was the International Association of Fire Chiefs organized?

30. Who were executed as arsonists in Rome after the Great Fire of 64 A.D.?

31. What city in Norway was almost totally destroyed by fire in 1624?

32. How did citizens of Cincinnati sound a fire alarm?

33. What was a "Brent Meister"?

34. Who were "Alcaldes de Carrio"?

35. What was Ben Franklin referring to when he said, "An ounce of prevention is worth a pound of cure"?

36. What was the purpose of the Fireman's Ball?

37. What Fire Department first used brass sliding poles?

38. What is the largest loss of life by fire ever recorded?

39. Who invented the first portable fire escape-ladder?

40. What Japanese city was engulfed by a fire storm in 1945?

History—Questions

41. Who rebuilt London after the Great Fire?

42. What city utilized the first practical fire engines?

43. What European city exported the best fire engines to colonial America?

44. What was the only city to have a fire horse college?

45. What city gave their horses annual vacations?

46. What was the name of the first volunteer fire company in America?

47. What were the first sliding poles made of?

48. What were firefighting soldiers from New York called in the Civil War?

49. What colonial city had the best Fire Prevention Program?

50. What did the first fire law of New Amsterdam forbid?

51. What man organized the first volunteer firemen in America?

52. What colonial city had the severest arson problem?

53. What was the "Rattle Watch"?

54. What city was the first to pay firefighters?

55. What caused most of the fires in colonial America?

56. What American city had the most severe fires in colonial times?

57. Who invented the first steam powered fire engine?

58. What was the name of the lithograph series by Currier & Ives that depicted fire scenes?

59. What city manufactured the most firefighting apparatus in the late 1800's?

60. What Fire Department responded to the first telegraph alarm?

61. Which American President watched a fight between volunteer fire companies in front of the White House?

62. How many years did Jamestown Virginia flourish before it was destroyed by fire?

63. What N.Y.C. Fire Co. first used a horse to pull it's rig?

64. What fire company was the first to wear red shirts, black pants and suspenders?

65. Oceanus Engine Co. paraded with a statue of what famous pioneer?

66. What were young men called who were interested in firefighting but who were not old enough to join a company?

67. In what city did the first woman firefighter work?

68. In what city did Stephen Decatur have a fire company named after him?

69. What city had the first firefighter's strike?

70. How long did N.Y.C.'s firefighter strike last?

71. In what month and year did N.Y.C. firefighters go on strike?

72. Who made the first American steam fire engine?

73. What was Ben Franklin's fire insurance company commonly called?

74. Which manufacturer was the first to make rubber hose?

75. In what year was the sprinkler system invented?

76. What was the month and year of the Watts riot and fires?

77. What European made the first steam fire engine?

78. What was the first fire department of ancient Rome called?

79. What was the first *paid* fire department of ancient Rome called?

80. Who started the first fire insurance company in London?

81. What fire department was the last to use a water tower?

82. What was the name of the volunteer firefighting group that was formed to protect England during World War II?

83. What contagious horse disease killed many fire department horses?

84. The first time in history that N.Y.C. requested mutual aid was in 1962. What was burning?

85. In ancient Rome, what were the people on fire watch called?

86. In ancient Rome, what were Aquarii?

87. In ancient Rome, who was the Prefectus Vigilum?

88. From what organization does the term "buff" come from?

89. Why were horses first used in the N.Y.C. Fire Department?

90. What was the first successful fire hose made of?

91. What city first used a fire alarm telegraph system?

92. What city first employed full-time firemen?

93. In what city did the two platoon system originate?

94. Who designed the first water tower?

95. What was the nickname for the volunteers on "Rattle Watch"?

96. How many leather buckets did Philadelphia order after it's first major fire?

97. What was the name of the first fire insurance company in America?

98. What city boasted the first fire insurance company?

99. Who built "Old Brass Backs"?

100. What did Benedict Arnold, Paul Revere and John Hancock have in common?

101. What Philadelphia fire company served with distinction in the Revolutionary War?

102. What did America's first fire regulation forbid?

103. What was the fine if a New Amsterdam house caught fire?

104. In what country was the first leather fire hose used?

105. Why weren't the alarm bells rung for the Great New Orleans Fire of 1788?

106. What English author wrote an essay called "American Volunteer Firemen"?

107. Who was the President when the British burned the White House?

108. What happened when the British torched an arsenal in Washington, D.C.?

109. What was a new engine called before it was "washed" by another company?

110. What did religious leaders in ancient times say should be done about fires?

111. What American city had the first recorded fire deaths?

112. Who invented the automatic sprinkler?

113. With what did the Insurance Company of North America mark insured buildings?

114. In what American city was the first fire apparatus manufactured?

115. What man was the first to successfully manufacture fire apparatus?

116. Who founded the first successful American fire insurance company?

117. What is the oldest fire insurance company in America?

118. Who contributed the first pumper to the Friendship Fire Co. of Alexandria, Va.?

119. The rank of foreman was the equivalent of what rank today?

120. What was considered the worst problem in American cities after the Revolutionary War?

121. In what American city were most of New England's fire apparatus built in the late 1700's?

122. Who manufactured the first horse drawn fire apparatus?

123. What was the first fire company to use a horse drawn fire apparatus?

124. What Engine Co. never once left it's firehouse?

125. What was the name of the first hose company in the U.S.?

126. What city was the first to carry water on the apparatus to fires?

127. What fire company was the first to put a bell and siren on their apparatus?

128. In what type occupancy was the first automatic fire sprinkler head system installed?

129. What city had the first black fire chief?

130. What city was the first to use an iron hulled fireboat?

131. What city was the first to install a high pressure water system?

132. Who invented the automatic fire sprinkler head?

133. What city formed the first flying squadron?

134. What department had the first radio system?

135.   What department consisted entirely of women?

136.   What was the greatest disgrace a volunteer company could suffer?

137.   What department had the first woman fire chief?

Answers

1. Augustus

2. It is a French word meaning "cover fire".

3. Thatched roofs

4. Between 1,000 and 2,000 men

5. A company of 50 women

6. A bellman

7. 1865

8. Soldiers

9. Beverly, Yorkshire (1726)

10. Boston (1677)

11. A fire engine manned by 12 men and a Chief in Boston

12. $10.00

13. Disassembling beds to bring them outside a burning building

14. Modern day salvage companies

15. Lack of discipline and refusal of the vollies to accept the new steam pumpers

16. They were afraid that the citizens would no longer respond and assist at fires

17. Cisterns

18. New York City 1914

19. North Carolina in 1914

20. 1863

21. Professor Moses Farmer of the Univ. of Maine

22. Dr. William Channing of Boston

23. China

24. Richard Newsham

25. Nicholas and Jan van der Heijden

26. The Radnor Fire Co. of Wayne, PA

27. 1949

28. King Richard I

29. Baltimore (1873)

30. Christians

31. Oslo

32. They beat a leather drum atop a carpenter's shop.

33. A fire warden in New Amsterdam

34. Fire wardens in New Orleans

35. Philadelphia's water supply

36. To raise money

37. Worcester, Mass.

38. 84,000 (bombing of Tokyo on March 9, 1945)

39. Abraham Wivell

40. Hiroshima

41. Christopher Wren

42. Boston

43. London

44. Detroit

45. Philadelphia

46. Mutual Fire Society

47. Wood

48. First New York Zouaves

49. Philadelphia

50. Wood or plaster chimneys

51. Peter Stuyvesant

52. Boston

53. Men who walked the streets of New Amsterdam at night carrying noisemakers which they sounded if they discovered a fire

54. Boston

55. Sparks from chimneys landing on roofs

56. Boston

57. George Braithwaite

58. "The Life of a Fireman"

59. Seneca Falls, N.Y.

60. Boston

61. John Tyler

62. One

63. Mutual Hook & Ladder #1

64. N.Y.C.'s Engine Co. 5 ("The Old Honey Bees")

65. Daniel Boone

66. Aides

67. N.Y.C.

68. Philadelphia

69. Boston

70. Approximately 5½ hours

71. November 1973

72. Paul Hodge

73. The Hand-in-Hand Company

74. Goodyear

75. 1806

76. August 1965

77. John Braithwaite

78. Familia Publica

79. Vigiles

80. Dr. Nicholas Barbon

81. Memphis, Tenn.

82. Auxiliary Fire Service

83. Epizootic disease

84. Brush

85. Nocturns

86. Firemen who formed a bucket brigade

87. The Chief

88. Buffalo Corps

89. To replace firemen sick from cholera

90. Leather and rivets

91. Boston

92. Cincinnati

93. Cleveland

94. Aaron Roberts

95. "The Prowlers"

96. 400

97. Friendly Society

98. Charleston, S.C.

99. Thomas Lote

100. All were volunteer firemen

101. Hibernia Co.

102. Wooden chimneys and thatched roofs

103. 25 florins

104. Holland in 1673

105. It was Good Friday and the priests refused to ring the bells

106. Charles Dickens

107. James Madison

108. An explosion killed over 100 British soldiers

109. A virgin

110. They considered fire to be God's vengeance and wanted to let them burn

111. Boston

112. John Carey (England, 1800)

113. A six pointed star

114. Philadelphia

115. Richard Mason

116. Benjamin Franklin

117. Philadelphia Contributionship

118. George Washington

119. Captain

120. Fire

121. Boston

122. Patrick Lyon

123. Good Will Fire Co.

124. Supply Engine Co. of N.Y.C.

125. Philadelphia Hose No. 1

126. Boston

127. Philadelphia Hose No. 1

128. A piano factory

129. Cambridge, Mass.

130. Boston

131. Rochester, N.Y.

132. Henry Parmalee

133. Detroit

134. Boston

135. Woodbine Ladies F.D. of Texas

136.   Being "washed" by another company

137.   Cedar Hill, Rhode Island

Apparatus

Most fire buffs have a keen interest in fire apparatus. They know precisely what type of rig a particular fire company has, and in many instances can tell you the type of truck the company had twenty or thirty years ago.

As a fire officer, I am concerned with the proper maintenance of the engine and equipment on board, the performance of the ladder, and the cleanliness of the truck. In short, will the truck get us to the fire and work properly when we get there. You could say in a nutshell, that the primary job of a fire officer is to insure that the apparatus and the fire fighters are ready to respond when the alarm comes in and will be able to perform their duties at the fire scene.

The Repairs and Transportation Unit of the

Apparatus

F.D.N.Y. is located in Long Island City, Queens. It is a huge garage encompassing an entire city block. At any one time there are more pumpers, trucks, and chiefs cars being worked on than most cities have in their entire fleet. Unfortunately, it is a sight most people interested in the fire service can not see. There is too much activity and a bit of danger to allow for visitor tours. However, if you are ever visiting New York, I'm sure no one will object if you stick your head in the door for a quick look.

Now let's see how much you know about apparatus.

These questions cover all three eras of the fire department as we know it. The hand pulled, the horse drawn and the motorized days are all included.

Apparatus

1. Who constructed one of the first fire engines in Augsburg, Germany in circa 1500?

2. What did Frederick Seagrave do before he made fire apparatus?

3. What type of vehicle was the first to have seat belts installed?

4. What was the name of the apparatus made by Gleason & Bailey?

5. What three men invented the first water tower?

6. Who built the last water tower?

7. What were the first names of the two LaFrance brothers?

8. Who patented a spring hoist device for raising aerial ladders?

9. What manufacturer built the first successful 100′ aerial ladder?

10. What was the nickname for early wooden aerial ladders?

11. What is the maximum age an "on line" apparatus should be?

12. Why should an apparatus not be parked directly behind a Seagrave rear mount aerial?

13. What is the actual working height of a 75′ Mack Tower Ladder?

14. What is the maximum road slope that a Tower Ladder should be operated on?

15. Who built the first brass fire engine?

16. What two manufacturers made a motorcycle and sidecar fire apparatus?

17. When did American LaFrance go out of business?

18. Who was the parent company of American LaFrance?

19. What fire department received the last American LaFrance apparatus made?

20. Who was the last president of American LaFrance?

21. For how many years was American LaFrance in business?

22. What is a pumper called in England?

23. What manufacturer made the pump for F.D.N.Y.'s Superpumper?

24. What symbol has always appeared on the doors of Nanuet, N.Y.'s fire apparatus?

25. What manufacturer first offered a full range of enclosed cabs?

26. Originally, what was the color of all American fire apparatus?

27. When did Ward LaFrance introduce diesel engines?

28. Who has the largest fleet of fire apparatus?

29. What is Canada's major fire equipment manufacturer?

30. Who sets the standards by which Canadian fire engines are built?

31. What is the largest water bomber in use today?

32. What did the Dennis brothers of England make before they made fire apparatus?

33. What department put the first "booster" apparatus in service?

34. Who developed the first steam fire engine in Ireland?

35. Who made the London Brigade Vertical?

36. What was the name of the first steam engine that Merryweather & Sons manufactured?

37. What is the name of the oldest steam fire engine still in existence?

38. Who built Washington, D.C.'s first steam fire engine?

39. Why did Elmira, N.Y. become a popular place to manufacture fire engines?

40. What Maryland manufacturer began making steam fire engines in 1858?

41. What manufacturer in Richmond, Va. produced steam fire engines beginning in the 1850's?

42. What two Philadelphia manufacturers began producing steam fire engines in the 1850's?

43. What N.Y.C. manufacturer began producing steam fire engines in the late 1850's?

44. What was the nickname of Henry Waterman's haywagon style fire engine?

45. Who made an engine known as "True Blue"?

46. Who made the London Brigade machine?

47. How many men did it take to crank up the Daniel Hayes aerial ladder?

48. What was Paul Hodge's steam fire engine called?

49. Who founded the LaFrance Co.?

50. Who built the Cincinnati F.D.'s "Uncle Joe Ross"?

51. In what country was the Laurin and Klement Company?

52.   Who built "Old Brass Backs"?

53.   What did Mack build before fire apparatus?

54.   Who made the first motorized turntable ladder?

55.   Who took over the Maxim Co.?

56.   Who took over the firm Hadley, Simpkin & Lott?

57.   Who produced the Deluge, Torrent & Fire King?

58.   For what manufacturer did Peter Pirsch first work?

59.   What company developed the first aerial platform for firefighting?

60.   What type of brake operates on an absence of air?

61.   Who built the steamer "Baltimore & Washington"?

62.   Who built the "Good Intent", "Hibernia", and "Mechanic"?

63.   Who built Europe's first self propelled steam fire engine?

64.   In what country is the Rosenbauer Company located?

65.   In what country is the Teudloff-Dittrich Co. located?

66.   In what country was the Scania-Vabis Company located?

67.   What was the name of England's first self-propelled steam fire engine?

68.   What manufacturer delivered the first aerial ladder truck to Washington, D.C.?

69.   What manufacturer has a bulldog insignia?

70.   In what year did Mack go into business?

71.   Which manufacturer's apparatus was distinguished by it's pump set in front of the radiator?

72.   Which manufacturer's engines were distinguished by a truncated front?

73.   In what year was the first Ahrens-Fox apparatus produced?

74.   How many Mack brothers were there?

75.   For what manufacturer did both Ahrens and Fox work for before forming their own company?

76. In 1915, what percent of F.D.N.Y.'s apparatus was motor driven?

77. From what country did the Steyr-Rosenbauer 46' aerial ladder come from?

78. In what country were the Air Raid Protection Vehicle and the Auxiliary Towing Vehicle developed?

79. Who makes the Simon Snorkel?

80. What is the rung spacing on most aerial ladders?

81. Who invented the diesel engine?

82. What country was the first to use diesel driven fire engines?

83. What company was the first to produce an electrically powered aerial?

84. What two men constructed the first water tower?

85. What company built water towers with either a telescopic pipe or a connecting pipe?

86. How many sections did the Scott-Uda aerial ladders have?

87. What was the official name of the first aerial ladder?

88. Who built the Dederick Aerial ladder?

89. What advantage did the Babcock aerial have over the Hayes aerial?

90. What company was the first to produce a spring assisted aerial ladder?

91. What does it mean to jack-knife a fire truck?

92. What is a buggy?

93. What is cavitation?

94. What is a governor on a fire truck?

95. Who made the Junior Aerial ladder?

96. What is a Pup boat?

97. What is a Nurse Tanker?

98. Who makes Snorkel apparatus?

99. Where is the manufacturer of Snorkel apparatus located?

100. What manufacturer makes the Squrt?

101. What was the first fireboat called?

102. What happened at a public demonstration of the Scott-Uda aerial ladder in 1875?

103. What is a quad?

104. What is a quintuple pumper?

105. What speed should a fire apparatus be able to maintain going up a 5% grade?

106. A fireman on a fireboat would know that a fathom is how many feet?

107. What time of day is the most hazardous time to drive a fire truck?

108. What does the term "slipping the clutch" mean?

109. Who designed the first mechanically raised aerial ladder?

110. What fire department operated the first steel aerial ladder?

111. What manufacturer designed the first all aluminum aerial ladder?

112. What is another word for a tormentor on an apparatus?

113. What company designed the first apparatus with an elevating platform?

114. What is the rotative force developed by an apparatus' engine called?

115. Where was the Lysander Button & Son fire apparatus built?

116. In what city did John Agnew build fire apparatus?

117. Who built the first American steam driven fire engine?

118. What New Hampshire manufacturer became the largest manufacturer of steam fire engines?

119. What two manufacturers merged to form the American LaFrance Co.?

120. In what year was America's first fireboat built?

121. Who made the first apparatus with suction hose and fittings?

122. Who patented America's first apparatus mounted aerial ladder?

123. What is the world's fastest fire truck?

124. What department was the first to become completely motorized?

125. What manufacturer was located in Minneapolis?

126. What manufacturer was located in Springfield, Mass.?

127. What manufacturer was located in St. Louis?

128. What manufacturer developed the centrifugal pump for firefighting?

129. What manufacturer was first to sell motorized apparatus on the West Coast?

130. What manufacturer was located in St. Paul?

131. What manufacturer was located in Vincennes, Indiana?

132. What department used the first all gasoline engine pumper?

133. What manufacturer made the first gasoline engine pumper?

134. What department bought the first all-powered aerial ladder truck?

135. What is the world's tallest elevating platform?

Answers

1.   Alton Plater

2.   He made wooden ladders

3.   A horse drawn fire apparatus

4.   Dederick Aerial Ladder

5.   Abner and Albert Greenleaf and John Logan

6.   American LaFrance

7.   Truckson and Asa

8.   Frederick Seagrave

9.   Fire Extinguisher Manufacturing Co. of Chicago

10.   Big stick

11.   25 years

12.   A firefighter will not be able to remove the portable ladders from the Seagrave

13.   70 feet

14.   15% grade

15.   Ctesibius of Greece

16.   Foamite-Childs Corp. and Indian Motorcycle Co.

17.   June 27, 1985

18.   Figgie International Co.

19.   Monroeville, Pennsylvania

20.   Patrick Clarke

21.   153 years

22.   An appliance

23.   DeLaval

Apparatus—Answers

24.   An indian chief's head

25.   American LaFrance

26.   Red-Brown

27.   1974

28.   The Federal forestry and lakes services

29.   Thibault AKA "Camions a Incendie Pierreville"

30.   The insurance companies

31.   Mars hydroplane

32.   Bicycles

33.   Cincinnati F.D.

34.   James Skelton

35.   Shand Mason

36.   The Deluge

37.   The Sutherland

38.   Poole and Hunt

39.   Many pump manufacturers were located there

40. Poole and Hunt

41. Eltenger and Edmond

42. Reaney & Neafie and Merrick & Sons

43. Lee & Larnard

44. "The Mankiller"

45. William Roberts

46. Merryweather Co.

47. Six

48. The Exterminator

49. Truckson Slocum LaFrance

50. Latta & Shank

51. Czechoslovakia

52. Thomas Lote

53. Buses

54. Magirus

55. Seagrave

56. Moses Merryweather

57.   Moses Merryweather

58.   Nicholas Pirsch Wagon & Carriage Plant

59.   Pitman Manufacturing Co.

60.   Maxi-brake

61.   Poole and Hunt

62.   Reaney and Neafie

63.   William Roberts

64.   Austria

65.   Sweden

66.   Hungary

67.   Merryweather Fire King

68.   Seagrave

69.   Mack

70.   1900

71.   Ahrens Fox

72.   John Christie Co.

73.   1911

74. Five

75. American LaFrance

76. 50%

77. Austria

78. England

79. Simon Engineering Dudley Co.

80. 14 inches

81. Rudolph Diesel

82. Bermuda

83. Webb Motor Co.

84. Abner Greenleaf and John Logan

85. Fire Extinguisher Mfg. Co. of Chicago

86. Eight

87. Hayes Hook & Ladder and Fire Escape Combined

88. Gleason & Bailey Works

89. The horses did not have to be unhitched to raise the ladder

90. Seagrave

91. Turn the tractor at an angle to the trailer

92. Chief Officer's vehicle

93. When air cavities are formed in a water pump

94. A device that controls the speed of the engine

95. Peter Pirsch & Sons

96. A small fireboat

97. A water supply truck which supplies pumpers

98. Snorkel Fire Equipment Co.

99. St. Joseph, Missouri

100. Snorkel Fire Equipment Co.

101. A floating engine

102. The ladder broke, killing three men

103. A combination pumper and ladder truck

104. Consists of aerial, portable ladders, fire pump, water tank and hose compartment

105. 35 MPH

106. Six feet

107. Dusk

108. Racing the engine with the clutch not completely engaged

109. Daniel D. Hayes

110. St. Paul F.D.

111. Peter Pirsch & Sons

112. Outrigger, jack or stabilizer

113. Pitman Manufacturing Co.

114. Torque

115. Waterford, N.Y.

116. Philadelphia

117. Latta & Shank

118. Amoskeag Co.

119. American Fire Engine Co. and LaFrance Fire Engine Co.

120. 1800

121. Sellers & Pennock

122. George Skinner

123. Jaguar XJ 12 "Chubb Firefighter"

124. Savannah, Georgia

125. Nott Fire Engine Co.

126. Knox Motor Co.

127. Robinson Fire Apparatus Manufacturing Co.

128. Seagrave

129. Seagrave

130. Waterous Engine Works Co.

131. Webb Motor Fire Apparatus Co.

132. Lansing, Michigan

133. Webb Motor Fire Apparatus Co.

134. Spokane, Washington

135. The Super Snorkel SS600 manufactured by Simon Engineering Dudley Ltd. of England has a working height of 202 feet.

Equipment

Give me a hose line and axe, a six-foot hook, a Halligan tool and men that know how to use them, and I will put out any fire. Well, almost any fire.

It is strange that in our high tech world, fire fighters use such simple tools; the same tools that firefighters used at the turn of the century. The reason is obvious; the tools were designed to fight fires and fire hasn't changed much in the last million years or so.

Every department has forcible entry tools. In New York City, we use the Halligan tool and axe primarily. The Halligan tool is the most versatile piece of equipment in the job. It can be used to force open most doors and can also be used to vent, open walls and floors, trim windows and pop automobile trunks open.

Equipment

At a working fire, while the engine company is stretching the line, the truck officer enters the fire building with two firefighters. One firefighter carries the Halligan tool and an axe, the other firefighter carries a sixfoot hook and a 2½ gallon water extinguisher. With these tools, the ladder company will force entry, search, and attempt to confine the fire until the engine company has charged the line and begins to move in.

Occasionally, when fire is out several windows on arrival, a young firefighter will think it is ridiculous to bring the 2½ gallon extinguisher. It is not ridiculous. Very often the extinguisher can be used to darken down a doorway and enable the forcible entry team to gain control of a door, closing it and keeping the fire out of the public hallway while the occupants escape down the stairs.

I have the utmost faith in the tools and equipment of the Fire Department because they are the best that money can buy and they are maintained properly by the firefighters and civilians in the department. I would not touch a piece of firefighting equipment that was not issued by the department. This was something I learned the hard way.

We responded to a hotel fire several blocks from the firehouse. I was driving that night and by the time I parked the rig, chocked the wheels, and entered the building, the rest of the troops had gone up in the only elevator. The fire was reported to be on the seventh floor and so I started walking up the stairs. I got to about the fifth floor when I heard on my handie-talkie that it was a mattress fire. Right in front of me, hanging on the wall was a soda-acid extinguisher. I took it to the seventh floor thinking it might come in handy.

As I got there, one of the men dragged the box spring out of a room and into the hallway. It was still smoldering and so I turned the extinguisher upside down to activate it. I heard a "bang". That was the last thing I heard. When I came to, I was sitting in a slumped over position several feet from where I had been. Someone was holding a handkerchief to my face while the other guys were getting me up onto my feet. The Chief's car was waiting when we got to the street and I was rushed to the hospital.

My lower lip had been torn open. I had a one inch gash on my chin and a puncture wound on my forehead. I had been hit by the bottom of the extinguisher which had been propelled like a missile at my face.

Hours later, back at the firehouse, we examined the extinguisher. Someone had painted it and in doing so, got paint on the threads. The result was, the cap did not fit tight. When I had inverted it, the cap blew off one way and the cylinder hit my face.

A short time later, the Fire Department banned all soda-acid extinguishers from use in New York City.

Incidentally, the plastic surgeon who worked on me did such a great job that the scars are barely visible. I am very thankful. It could have been a whole lot worse.

Equipment

Here are some questions on equipment used in days gone by as well as questions on modern tools and equipment—everything from the fire pail to the Hurst Tool.

Equipment

1. What is the color of an oxygen cylinder?

2. What is the color of an acetylene cylinder?

3. What is the maximum psi that acetylene can be safely used at?

4. What does marlinspike seamanship refer to?

5. What is the very end of a rope called?

6. What is a bight on a rope?

7. What knot forms an eye which cannot slip and which is easily taken apart?

8. What % of a rope's strength is retained after splicing?

9. On what class fire can a plain water extinguisher be used?

10. What is the main advantage to using a carbon dioxide extinguisher as opposed to another type?

11. Multipurpose dry chemical extinguishers have a base of what chemical?

12. What is the danger of using a Class A extinguisher on a Class B fire?

13. What is the danger of using a Class A extinguisher on a Class C fire?

14. On what type fire would a dry powder extinguisher be used?

15. What two chemicals are in a soda-acid extinguisher?

16. On what class fire is foam used?

17. In what year was the manufacture of all inverting type extinguishers discontinued?

18. How do you determine if a carbon dioxide extinguisher is fully charged?

19. What unique shape does a fire bucket have?

20. What is the color of a fire bucket?

21. Fire blankets can be used to smother flames in what class fire?

22. What is the ideal angle th t a portable ladder should be placed ιgainst a fire building?

23. What company made the first rubber hose?

24. What fire department was the first to use the rubber hose?

25. In what year was rubber hose first used in the fire service?

26. What fire department was the first to use compressed air tools?

27. What was the first mask designed for fire fighters called?

28. What was the first American company to design a self-contained breathing apparatus?

29. What is the name of a wrench used to tighten hose butts?

30. Who built the first American extinguisher?

31. What piece of equipment did the Uncinarius of ancient Rome carry?

32. When cutting metal with an oxy-acetylene torch, what temperature must the metal be raised to?

33. How long should an explosimeter warm up to get the most accurate result?

34. What does it indicate if the needle of the explosimeter is at 80?

35. What three filters are used with the explosimeter?

36. What is an eductor used for?

37. What is the principle by which an eductor works?

38. How often should hose be tested under pressure?

39. If cotton hose is not dried properly, what will form on it?

40. When loading hose, what is the inside of the coupling checked for?

41. What government agency approves masks?

42.   Life nets are not practical to use for persons jumping above what floor?

43.   What tool is known as the Jaws of Life?

44.   What is the nob or knob?

45.   What do rubber boots contain that will allow electric current to pass through them?

46.   What chemical should be put into an extinguisher to keep it from freezing?

47.   Why should salt not be used in an extinguisher to keep it from freezing?

48.   Why should glycol not be used in an extinguisher?

49.   What Dow Chemical product is used to fight forest fires?

50.   On a ladder, what type plate connects the rails of the beams and supports the rungs?

51.   What is a halyard?

52.   A dog or pawl are names for what?

53.   What does discoloration of an aluminum ladder usually mean?

54. What is the climbing angle for a portable ladder?

55. What two things do aluminum ladders conduct?

56. What is the maximum safe load for a 25′ extension ladder?

57. Which is easier to maintain—wooden or aluminum ladders?

58. What is a "K" tool used for?

59. What is a key tool used for?

60. What is the weight of one charged length of 1¾ inch hose?

61. How long was the MSA Chemox oxygen mask rated for?

62. What does MSA stand for?

63. In what city is the MSA Co. located?

64. What is the weight of a charged 40′ length of leather hose?

65. What is the oldest modern portable fire extinguisher?

66. What type extinguisher did the British Home Office distribute to homeowners during WW2?

67. What is the length of a 6' hook?

68. What gas besides air can be used to charge a water extinguisher?

69. Who invented the fire hose?

70. What was the first fire hose made of?

71. In what year was the Manby fire extinguisher invented?

72. In what year was the Wivell fire escape invented?

73. What is the name of the valve just inside a siamese inlet?

74. What does exposure to light do to a rubber turnout coat?

75. What will drying a work glove on a radiator or stove do to it?

76. What instrument is used to test for the presence of flammable vapors and gases?

77. Who developed bromochloromethane as an extinguishing agent?

78. What is bromotrifluoromethane commonly known as?

79. What manufacturer developed Halon 1301?

80. Who had the idea to sew an iron band into a fire helmet to reduce warping?

81. What manufacturer of fire helmets, founded in 1828, is still in existence?

82. What is the shelf life of pre-mixed aqueous film forming foam?

83. What are protein foams made of?

84. What is the approximate cost of a complete Hurst Tool?

85. How much force in pounds does the Hurst Tool have at the tips of the jaws?

86. How many firefighters are needed to operate the Hurst Tool?

87. How long are the hoses on the Hurst Tool?

88. What is wrong with fiberglass fire helmets?

89. Why do firefighters occasionally wear their helmets backwards?

90. What would cause the brim of a leather helmet to bend down at a right angle?

91. What type of device is attached to a hose line and lowered to inaccessible positions?

92. What is a chuck?

93. What is a church raise?

94.   What is a churn valve?

95.   What is a cistern?

96.   Describe a claw tool.

97.   What is a combination nozzle?

98.   What is a "baby bangor"?

99.   Where was the bangor ladder developed?

100. Describe a bayonet nozzle.

101. What are bunker clothes?

102. What is a "can"?

103. What is a constant flow nozzle?

104. What does it mean to dog a piece of equipment?

105. What is a double female?

106. What is a double lay?

107. What is a flat load?

108. What manufacturer originally developed foamite?

109. What is the name of the sealing ring in a coupling, which provides a water tight connection?

110. What piece of rope is used to secure hose?

111. What tool is used to open fire hydrants?

112. How many gallons of water does an Indian pump hold?

113. Who makes Indian pumps?

114. What is the usual diameter of the non-collapsible hard suction hose?

115. What is the insignia plaque on the front of a firefighters helmet called?

116. What is the heel of a ladder?

117. What is a higbee cut?

118. What device facilitates pulling hose over window sills?

119. What is a joker?

120. What is a kelly tool?

121. What is a sharp bend in a hose called?

122. What is a ladder spur?

123. What is a reverse lay?

124. What is a leg lock?

125. How long is a length of fire hose?

126. What tool fires a cord to trapped persons?

127. What is a jumping sheet?

128. What were life nets originally made of?

129. What manufacturer makes light water?

130. On what type fire is light water used?

131. What is a mall?

132. What is a mattock?

133. In what two solutions is mechanical foam concentrate made?

134. What does "over the air" mean?

135. What is another name for an imaginary or dummy fire alarm box?

136. What is a playpipe?

137. What is a johnny pump?

138. What is a fire plug (in modern times)?

139. What is a pneolator?

140. What are sliding poles usually made of?

141. What is a pony extinguisher?

142. What is potable water?

143. Who makes Purple K?

144. What is a staypole?

145. What is the advantage of a trussed ladder over a solid beam ladder?

146. What is a water bag?

147. What is the term for the impact energy caused by shutting down nozzles too quickly?

148. Who invented the scaling ladder?

149. What type of ray does acetylene burning produce?

150. When cutting with acetylene, where should the hose be?

151. What type of iron is used to move an elevator car away from a trapped person?

152. What piece of rescue equipment gives a mechanical advantage in lifting heavy objects?

153. What is a snatch block used for?

154. What is it called to wrap a piece of twine around a hook to keep a rope from jumping out?

155. What happens to a hook in a block and tackle before it fails?

156. On what piece of equipment do you find a becket, thimble, sheave and swallow?

157. What is the liquid in an acetylene cylinder called?

158. What is the color of oxygen hose?

159. What is the color of acetylene hose?

160. What is the full name of the company that manufactures Stang nozzles?

161. Who invented the earliest known water pump?

162. How many gallons of water did the leather water bucket of colonial days hold?

163. If a saw blade is turning at 6,000 rpm., how fast in mph. is it going?

164. What will happen to a saw blade if a straight kerf is not maintained?

165. Which Partner saw blade is used to cut through tar roofs and wood flooring?

166. Which Partner saw blade cuts concrete?

167. Which Partner saw blade cuts through steel?

168. How much fuel does the Partner saw hold?

169. How long will a Partner saw run when starting with a full tank?

170. When cutting a roof, how deep should the cut be?

171. Why should the engine of a Partner saw not be gunned while moving from one location to another?

172. What will happen if the V belt on a Partner saw is too tight?

173. What piece of equipment is composed of the following parts: cylinder, yoke, schrader plug, aspirator?

174. What type of wood is an axe handle usually made of?

175. What type of wood is a hook handle made of?

176. How many men did it take to raise a 75′ extension ladder?

177. What is the world's most powerful fire appliance?

178. What does difficulty in calibrating an ion chamber indicate?

179. When a nozzle is closed, how does the nozzle pressure compare to the pumper pressure?

180. Is more damage done to a charged or uncharged line, if a vehicle goes over it?

181. Why shouldn't brass polish be used on hose couplings?

182. How many firefighters can safely climb a 35' ladder at one time?

183. What is another name for a Pompier ladder?

184. In what country is the Partner saw made?

185. What does wetting a splice in a rope do?

186. By what percent does a clove hitch or bowline knot reduce a rope's strength?

187. In what year was the first leather fire helmet made?

188. Will a self contained breathing apparatus protect against the inhalation of airborne radioactive particles?

Answers

1. Green

2. Red

3. 15 psi.

4. The use of rope

5. The bitter end

6. A loop formed by turning the rope back on itself

7. A bowline

8. 85-95%

9.   Class A

10. It leaves no residue

11. Monoammonium phosphate

12. Flareup of the fire

13. The operator could receive a shock

14. Combustible metal

15. Sodium bicarbonate and sulfuric acid

16. Class B

17. 1969

18. Weigh the extinguisher

19. It has a rounded or pointed bottom

20. Red

21. Class A

22. 60 degrees

23. Goodyear

24. Cincinnati F.D.

25. 1871

26. San Francisco F.D.

27. The Aldini

28. W.E. Gibbs

29. Spanner

30. Joseph Jynks

31. A hook

32. Its ignition temperature

33. Five minutes

34. The concentration of flammable gas is approaching the lower explosive limits

35. Cotton, inhibitor, charcoal

36. Pumping out flooded areas

37. Venturi principle

38. Annually

39. Mildew, mold

40. The washer

41. U.S. Bureau of Mines

42. The 4th floor

43.   Hurst tool

44.   The nozzle

45.   Carbon black

46.   Calcium chloride

47.   It is too corrosive

48.   Glycol can ignite under certain conditions

49.   Gelgard

50.   Gusset plates

51.   A rope used to extend the fly section of a ladder

52.   Ladder locks

53.   Loss of structural strength

54.   60-75 degrees

55.   Heat and electricity

56.   500 pounds

57.   Aluminum

58.   Forcible entry, pulling lock cylinders

59.   Forcible entry

60.   78 pounds

61.   Three quarters of an hour

62.   Mine Safety Appliance

63.   Pittsburgh

64.   150 pounds

65.   Soda-acid

66.   Stirrup pump

67.   6½ feet

68.   Nitrogen

69.   Jan van der Heyden

70.   Leather

71.   1816

72.   1830

73.   Clapper valve

74.   Shortens its life

75.   Damage the fibers and shorten the life of the glove

76.   Explosimeter

77. German World War 2 scientists

78. Halon 1301

79. E. I. DuPont de Nemours & Co.

80. Matthew DuBois

81. Cairns & Bros.

82. 6 months

83. Hydrolized protein, polyvalent metallic salt, organic solvents

84. $11,000

85. 10,000 pounds

86. One

87. 16 feet

88. No resilience

89. The brim protects against heat

90. Improper storage

91. Distributor

92. Portable hydrant carried on the apparatus

93. Raising a ladder with top supported by four guy lines

94. Pressure relief valve

95. Water storage container usually below grade

96. It has a hook and fulcrum at one end and a claw at the other

97. An all purpose nozzle providing either solid stream or fog

98. Small extension ladder without ropes or pulleys

99. Bangor, Maine

100. Small volume spray applicator with piercing head

101. Nightime turnout clothes

102. A fire extinguisher

103. A fog nozzle which provides a constant flow regardless of spray pattern

104. To secure it

105. A coupling with two female swivel couplings

106. A company lays out two lines simultaneously

107. Hose laid flat in the hose bed

108. Foamite-Childs Corp.

109. Gasket

110. Hose strap

111. Hydrant wrench

112. Five

113. D.B. Smith Co. of Utica, N.Y.

114. 4½"

115. Hat shield or hat front or frontpiece

116. The base

117. The outside thread of a coupling is removed to prevent crossing of threads

118. Hose roller

119. The bell in the firehouse upon which alarms are received

120. Forcible entry tool with adz at the top and a fork at the bottom

121. Kink

122. A levelling device on the ladder heel

123. A method of laying hose from an apparatus

124. A means of securing oneself to a ladder

125. 50′

126. Life gun

127. Life net

128. Rope

129. Minnesota Mining & Manufacturing Co.

130. Flammable liquids

131. A sledge hammer

132. A hand tool for digging and constructing fire lines

133. 3% and 6%

134. Via radio

135. Phantom box

136. A tapered hosepipe with handles

137. Hydrant

138. Hydrant

Equipment—Answers

139. Resuscitator

140. Brass

141. Extinguisher with less than 2½ gallons

142. Water fit for humans to drink

143. Ansul Co. of Marinette, Wisconsin

144. Pole used to brace a ladder

145. Trussed ladder is stronger

146. Containers of water on planes which are dumped on forest fires

147. Water hammer

148. Chris Hoell, a St. Louis fireman

149. Ultraviolet rays

150. Behind the operator

151. Z iron

152. Block & Tackle

153. To achieve a change of direction in pulling

154. Mousing

155. It will start to spread

156. A straight block or block & tackle

157. Acetone

158. Green

159. Red

160. John W. Stang Manufacturing Co. Inc.

161. Ctesibius of Alexandria in 200 B.C.

162. 3 gallons

163. 215 mph.

164. It will bind

165. Carbide tip

166. Silicon carbide

167. Aluminum oxide

168. One quart

169. ½ hour - 1 hour

170. Only as deep as necessary to remove roof boards

171. It will cause the blade to spin

172. The blade will spin while the saw is idling

173. Resuscitator

174. Hickory

175. Ash

176. Nine

177. The 860-HP Oshkosh fire truck

178. Weak batteries

179. Approximately the same

180. Uncharged

181. It cakes and prevents swivel movement

182. Three

183. Scaling ladder

184. Sweden

185. Makes it stronger than the original rope

186. 60%

187. 1740

188. Yes

Famous Fires and Disasters

Since prehistoric times, fires have ravaged the earth. Fire is a part of nature which must be controlled by man and it is only in the past few centuries that any serious attempts have been made to accomplish this. From the burning of Rome to the Great London and Chicago fires, right up to the costly 1985 brush fires in California and Florida, entire communities have succumbed to the onslaught of the Red Devil. Even with modern equipment and trained personnel, fire departments are hard pressed at times to keep fire from destroying our lives.

Fires, whether accidental or arson-related, will always be with us, and there are more coming in the future which will rival the big ones of the past. Many could be prevented with common sense but others will happen regardless of precautions. Lightning will strike the forest, the sun will reflect off a shiny object and start brush on fire, and rodents will chew through electrical cable and set a building on fire.

Other fires will be attributed to our standard of living. Oil burners and kitchen stoves malfunction, appliances overheat, and various chemicals react with one another and ignite. And of course there are always man-made fires set for revenge, profit or just kicks.

Finally, looming on the horizon is the possibility of a nuclear explosion occurring in a major city. This would be the ultimate nightmare for citizens and firefighter alike. With the entire city in flames, fire equipment destroyed, and most firefighters killed, there would be no hope for anyone left alive. This doomsday fire would then top the list of famous fires. Let us hope and pray it never happens.

In doing research for this book, I was surprised at the number of fires that the early settlers of our country had to contend with. I lost count of the number of times that Boston burned. Although I've lived in New York all of my life, I never knew of the severity of the fires in New York City during the eighteenth and nineteenth centuries. Entire sections of the city were destroyed on several occasions.

We usually think that major fires and disasters occur in other countries. We hear of terrible earthquakes, mud slides, typhoons, explosions, etc.

happening in places like Pakistan, China, and Italy. But unfortunately, just as many disasters occur in the United States as in other countries. San Francisco and Anchorage have been rocked by earthquakes; tornadoes touch down in the Midwest destroying entire neighborhoods; hurricanes hit the East Coast every year; and every so often a city will be almost destroyed by fire as Lynn, Massachusetts, was in 1981.

Fires and disasters will continue to occur until the end of time. The fire department is the first line of defense against these tragedies and we must always be ready to deal with them.

But right now, let's not worry about the future but rather think of the past. Here are questions about some of the worst fires and disasters in history.

Famous Fires and Disasters

1. Where did the biggest fire during World War 2 occur?

2. Who was the radio announcer who broadcast the Hindenburgh disaster?

3. From what city was the Morro Castle sailing when it caught fire?

4. In what part of the Morro Castle did the fire start?

5. What happened at LeMans on June 11, 1955?

6. What two types of aircraft collided over Brooklyn, N.Y. in 1960?

7. What Colorado "boom town" burned to the ground in 1896?

8. What movie star was killed in the Cocoanut Grove fire?

9. Where did the Hindenburg disaster occur?

10. What type airplane crashed into the Empire State Building?

11. Where did the world's worst train wreck occur?

12. Where did the world's worst aircraft disaster occur?

13. Where did the world's worst mine disaster occur?

14. What was the worst single airplane disaster?

15. How many Chicago firefighters were killed at the 1893 Columbian Exposition fire?

16. Where did America's worst schoolhouse fire take place?

17. How many firefighters were killed at the Chicago stockyard fire in 1910?

18. How many firefighters were killed in the Philadelphia leather factory fire in 1910?

19. What gas killed most of the patients at the Cleveland clinic fire of 1929?

20. Why was the death toll so high in the Ohio State Penitentiary fire?

21. What entertainer was mistakenly declared missing at the Kentucky nightclub fire in 1977?

22. Between what two cities did the Morro Castle sail?

23. With what did the waiters at the Cocoanut Grove try to fight the fire?

24. What famous clown was performing the day of the Hartford circus fire?

25. How many convicts were killed in the Ohio State Penitentiary fire?

26. How did the Iroquois theatre fire start?

27. On what holiday weekend did the Cocoanut Grove fire occur?

28. What kind of gas exploded at the East Ohio Gas Co. in Cleveland killing 130?

29. In what famous fire and explosion was every firefighter killed and every piece of equipment destroyed?

30. What chemical exploded in the Texas City, Texas fire?

31. What was the name of the ship that exploded in Texas City, Texas?

32. In what country was the greatest gas fire in history?

33. In what country were 1670 people killed in a theatre fire?

34. What country had the most deaths in a hotel fire?

35. In what city were the most people killed in a circus fire?

36. How many buildings were destroyed in the Sacramento fire of 1852?

37. How many volunteer firemen were killed in the Jennings clothing store fire in 1854?

38. Where did the first recorded fire in America take place?

39. Where did America's first arson fire take place?

40. What was the name of the Dutch ship that burned in N.Y. harbor in 1613?

41. At what fire did 31 floors of a fireproof building burn?

42. What happened at the Torrey Canyon disaster?

43. What ship blew up in the Mississippi River killing over 1500 Union soldiers?

44. How many lives were lost in the General Slocum fire?

45. Who was Capt. William H. Van Schaick?

46. What was the last theater fire disaster in the U.S.?

47. How many people perished in the Iroquois theatre?

48. How many people were killed in the San Francisco earthquake and fire?

49. What was the seismograph reading at the San Francisco earthquake?

50. What was the year of the Great Chicago fire?

51. In what year was the San Francisco earthquake?

52. Where was the Triangle Shirtwaist factory located?

53. What started a fire on July 4, 1866 in Portland, Maine that destroyed 1500 buildings?

54. Whose cow kicked over the kerosene lamp starting the Great Chicago fire?

55. Who owned the museum in which N.Y.C. volunteer firemen fought their last major fire?

56. What is the Great Boston fire also called?

57. Approximately what percent of Chicago was destroyed in the Great Chicago fire?

58. Of all the great fires in America, which one generated the most heat?

59. What was the loss in dollars of the Bel-Air Brentwood fire in 1961?

60. What was the year of the Great London fire?

61. How many people died in the Great London fire?

62. In what year did Rome burn while the Emperor fiddled?

63. What did the mayor of London do when he was awakened to see the Great fire?

64. How many Philadelphia volunteer firemen were killed at a fire on May 9, 1791?

65. Name the three astronauts who were killed when their spacecraft caught fire?

Answers

1. Dresden, Germany

2. Herb Morrison

3. Havana

4. The library

5. A Mercedes left the track, exploded and killed 82

6. Douglas DC8 and Lockheed Super Constellation

7. Cripple Creek

8. Buck Jones

Famous Fires and Disasters—Answers

9.   Lakehurst, N.J.

10.   B-25

11.   Modane, France (1917 - 543 killed)

12.   Tenerife, Canary Islands (1977 - 581 killed)

13.   Manchuria (1942 - 1549 killed)

14.   Japan Air Lines Boeing 747 in Japan (1985 - 520 killed)

15.   12

16.   Collinwood, Ohio

17.   21

18.   14

19.   Nitrogen dioxide from burning x-rays

20.   Guards refused to open cells

21.   John Davidson

22.   New York City and Havana

23.   Seltzer bottles

24.   Emmett Kelly

25.   320

26.   Stage lamp ignited backstage scenery

27.   Thanksgiving

28.   Liquefied methane gas

29.   Texas City, Texas fire and explosion

30.   Ammonium nitrate

31.   SS Grandcamp

32.   Algeria

33.   China

34.   South Korea

35.   Hartford, Connecticut

36.   7,000

37.   11

38.   Jamestown, Virginia

39.   Plymouth, Mass.

40.   Tiger

41.   Andraus building fire in Sao Paolo, Brazil

Famous Fires and Disasters—Answers

42. 118,000 tons of crude oil went into the ocean, polluting the English and French coasts

43. The Sultana

44. 1,030

45. Captain of the General Slocum

46. Iroquois theatre

47. 602

48. 674

49. 8.25

50. 1871

51. 1906

52. New York City

53. A firecracker

54. Mrs. Kate O'Leary

55. Phineas T. Barnum

56. The Epizootic fire

57. Approximately 33⅓%

58.   The Great Chicago fire

59.   25 million

60.   1666

61.   6

62.   64 A.D.

63.   He went back to sleep

64.   12

65.   Virgil Grissom, Edward White, Roger Chaffee

Famous Firefighters

In every fire company, there are one or two individuals who excell in their firefighting abilities. For some it comes naturally, but for others, much hard work is necessary. These individuals gain a reputation throughout the department. They are known for their aggressiveness at fires, their skill with tools and their ingenuity during emergencies. Younger firefighters admire and emulate them and other companies try to persuade them to transfer. Unfortunately, in most cases their names are not known outside of the department to which they belong. Chicago could have a superhuman firefighter known throughout the job, but he would be unknown in Los Angeles or Dallas. Atlanta's most decorated firefighters would be unknown in St. Louis or San Diego.

These heroes are rewarded by the satisfying feeling of a job well done. They are also awarded citations and medals for some of the more spectacular feats that they perform. The general public may occasionally read about these heroic deeds in the newspapers but only those in the fire service can fully appreciate the accomplishments of these firefighters.

There has only been one man in the history of the F.D.N.Y. to have his badge retired. Although I never worked with him at a fire, I did meet him when he was working a light duty job just prior to his retirement. His name is the answer to one of the questions below.

As for exceptional firefighters I have worked with, two men come to mind immediately. I considered one of these men a "one-man fire department." At a roof operation, he would cut a hole with the saw, then grab a halligan tool to pull back the tar, pry off the roof boards and push down the ceiling below. Then using the saw again, he would widen the hole and continue this routine until the engine company below had moved in and extinguished the fire. I used to joke that he rarely did any overhauling. He was usually on his way to the hospital suffering from smoke inhalation, exhaustion and conjunctivitis.

The other man knew his job thoroughly and in addition was a tremendous worker around the firehouse. When he was on duty, he would spend hours working on the apparatus. He would clean, wax, grease, oil and gas it. Then he would get out a paint brush and paint the diamond plate or the wheels. The tools would be cleaned and lightly oiled, the masks checked, the saw started and all other equipment inspected. On top of this, he was a pleasant man to work with and was always willing to help a new man learn the job.

Everyone knows firefighters like these men. Here are some questions about other famous firefighters who have served. There may be a ringer or two in the group, but see how many you know.

Famous Firefighters

1. What U.S. President was an honorary member of the Friendship fire company in Alexandria, Va.?

2. What fireman of the Union Co. in Lancaster, Pa. became President of the United States?

3. What fireman from the Eagle Hose Co. in Buffalo became President of the United States?

4. What fireman designed America's first fire helmet?

109

5.   What fireman died in the Boston Massacre in 1770?

6.   What American firefighter was the first to have the title "Chief"?

7.   What firefighter serving in the Revolutionary War was known as the "boy colonel"?

8.   What Philadelphia fireman captured the frigate Randolph during the Revolutionary War?

9.   What N.Y.C. volunteer later had a fireboat named after him?

10.   What Philadelphia fireman organized America's first hose company?

11.   What fireman bought the first horse to pull a fire truck?

12.   What fireman designed the first leather helmet?

13.   Who was the first woman volunteer firefighter in America?

14.   Who was Pittsburgh's first woman firefighter?

15.   Who was considered the most famous woman firefighter in America in the 1800's?

16. What N.Y.C. mayor known as "Boss", was a fireman?

17. What famous printer, who later formed a partnership with James Ives, was a volunteer fireman?

18. What volunteer fireman later became president of Seamen's Savings Bank?

19. What fireman was portrayed in the play "A Glance at New York"?

20. Who was the last chief of N.Y.C.'s volunteer fire department?

21. What firefighter became a U.S. Senator and was killed in a duel?

22. What fireman invented the quick hitch collar and hame?

23. What firefighter invented the sliding pole?

24. What fireman had the idea to pre-heat the water in the steamer before responding?

25. What firefighter invented an improved metal collar which adjusted to a horse's neck?

26. What fire captain invented the sliding pole?

27. What fire captain invented the first brass sliding pole?

28. What Chicago firefighter turned in the first alarm for the Great fire?

29. What Englishman was Edinburgh's first chief, wrote a handbook on fire department operations, and died in the line of duty?

30. Who was the first chief of the Boston F.D.?

31. Who first suggested that hydraulic platforms be used for firefighting?

32. Who was the first chief of the Salt Lake City F.D.?

33. Who invented the automatic horse cover?

34. What type fire does Red Adair fight?

35. Who was America's first woman fire chief?

36. What two tools did Chief Hugh Halligan invent?

37. What firefighter war hero organized Chicago's ambulance service?

38. What grandmother was once chief of the Woodbine Ladies F.D. of Texas?

39. What Los Angeles battalion chief invented the modern fog nozzle?

40. What firefighter from Auburn, N.Y. won a Bronze Star on Iwo Jima and later became editor of "Fire Engineering" magazine?

41. What Assistant Chief of the Cincinnati F.D. helped found the Ahrens-Fox Co.?

42. Who extinguished the greatest gas fire in history?

43. What firefighter killed a bengal tiger at the Barnum museum fire?

44. Who was the first black fire chief in America?

45. Who was the first St. Louis firefighter to die in the line of duty?

46. Who was in charge of N.Y.C.'s first engine company?

47. What Chicago Fire Commissioner coined the term "snorkel"?

48. Who invented the Dahill Hoist?

49. What firefighter founded Cairns & Bros.?

50. What Kansas City chief designed a hydraulic water tower?

51. What N.Y.C. Fire Commissioner swore in his own son as a probationary fireman?

52. What British firefighter of the early 1800's had a fireboat named for him?

53. What N.Y.C. firefighter was the winner of the first N.Y.C. Marathon?

54. Who authored "20,000 alarms?"

55. Who was London England's first fire chief?

56. What Golden Gloves champion became president of N.Y.C.'s fire union?

57. Who is the only man in N.Y.C.'s history to have his badge retired?

58. London England's first fire chief died in the line of duty. What killed him?

59. Who was the second chief of the London Fire Brigade?

60. What union leader led N.Y.C. firefighters on strike?

61. Who formed the first volunteer fire company in Philadelphia?

62. What retired Chicago firefighter was on "fire watch" at the Iroquois theatre?

63. Who was America's first paid firefighting officer?

64. What St. Louis firefighter invented the scaling ladder?

65. Where did Peter Pirsch work as a volunteer fireman?

Answers

1. George Washington

2. James Buchanan

3. Millard Fillmore

4. Jacob Turck

5. Samuel Maverick

6. Jacob Stoutenburgh

7. Walter Stewart

Famous Firefighters—Answers

8. Nicholas Biddle

9. Zophar Mills

10. Reuben Haines

11. Gabriel Disosway

12. Henry T. Gratacap

13. Molly Williams

14. Marina Betts

15. Lillie Hitchcock Coit

16. William "Boss" Tweed

17. Nathaniel Currier

18. William Macey

19. Mose Humphreys

20. John Decker

21. David Broderick

22. Charles Berry

23. Daniel Lawler

24. William Gleason

25.   John Freyvogel

26.   Capt. David Kenyon of Chicago F.D. in 1878

27.   Capt. Charles Allen

28.   Joseph Lauf

29.   James Braidwood

30.   Thomas Atkins

31.   Capt. Quinn of Chicago F.D.

32.   George M. Ottinger

33.   Chief George Hale of Kansas City F.D.

34.   Oil well fires

35.   Nancy Holst

36.   Halligan tool, Halligan hook

37.   Joseph McCarthy

38.   Verlie Gunter

39.   Glenn Griswold

40.   Donald O'Brien

Famous Firefighters—Answers

41. Charles Fox

42. Red Adair

43. John Dedham, a N.Y.C. volunteer

44. Patrick Raymond

45. Capt. Thomas Targee

46. Peter Rutger

47. Robert Quinn

48. Chief Edward Dahill of New Bedford, Mass.

49. Henry Gratacap

50. George Hale

51. Edward Thompson

52. James Braidwood

53. Gary Murchke

54. Richard Hamilton of N.Y.C.'s Rescue 2

55. James Braidwood

56. Michael Maye

Famous Firefighters—Answers

57. Richard Hamilton

58. A wall collapsed

59. Captain Sire Eyre Massey Shaw

60. Richard Vizzini

61. Benjamin Franklin (Union Fire Co.)

62. William Sallers

63. Thomas Atkins

64. Chris Hoell

65. Kenosha, Wisconsin

Tactics

There are thousands of fire departments in the United States, both professional and volunteer. Although each department will fight a fire in basically the same manner, there can be variations in how the job is done. These would depend on the experience of the force, training, leadership, policy, tradition and espirit de corps.

Over the years, a department will perfect operations at a particular type of fire with which it must deal constantly. A fire in an occupied residence in the middle of the night can become routine in the sense that the firefighters arriving on the scene have faced this type of fire before. They know what to expect and what has to be done to evacuate the building and extinguish the fire.

This is not to say that the job becomes easier with experience. On the contrary, in one way it becomes more difficult. Inexperienced firefighters arriving at such a scene might think they are doing a good job merely by putting water on the flames. Veteran firefighters would see the big picture and realize all that must be done. Water must be put on the seat of the fire, doors must be forced, apartments searched for both occupants and fire, and ceilings and walls must be opened to check for fire extension. All must be done in a quick and efficient manner in a hostile environment of smoke and high heat. The reward for knowing how to do the job is satisfaction at having completed the task and probably a king-sized headache.

The questions and answers in this section are apt to be controversial. They are based mostly on personal experience with the exception of the questions on forest fires. It is the way things are done in N.Y.C. and presumably in most large cities. However, members of some departments may not agree with all of the answers.

At any rate, years ago while fighting my first few fires as a "johnnie", you would have had a tough time convincing me that what we were doing was tactical. My only concern was trying to keep up with my officer and staying alive at the same time. It was only after a year or so of heavy fire duty that I realized that there was a method to what seemed like madness. I liked the job from the start, but it became even more enjoyable when I learned all of my responsibilities and duties at a fire.

124

Tactics

Here are some questions on tactics. See if you would fight the fire the same way I would.

Tactics

1. At a cellar fire in a multiple dwelling with an interior and exterior staircase, where should the first hand line go?

2. What is the most efficient method of extinguishing a gas fire?

3. A five story tenement has no fire escape on the front. How many apartments are there in the building?

4. What are the four ways to extinguish a fire?

5. What would probably happen if a solid stream of water hit molten lead?

6. Compared to a conventional home, how much faster does a mobile home burn?

7. What is the number one cause of fire in a mobile home?

8. Which terminal of a car battery should be disconnected first?

9. What does a water curtain protect?

10. What areas of a building are most susceptible to a backdraft?

11. If caught in a backdraft, what should a firefighter do?

12. Before ascending or descending a gooseneck ladder on a fire escape, what should a firefighter do?

13. For safety reasons, what should the maximum number of firefighters on a fire escape landing be?

14. In what room do occupants of private dwellings and high rise buildings seek refuge?

15. Which is more stable — a bearing wall or non-bearing wall?

16. Which is more stable: a bearing wall in a

five story building or a bearing wall in a
two story building?

17. What is the maximum distance the turntable of an aerial ladder should be placed
from a wall of a fire building?

18. Aerial ladders are extended, raised and
rotated. What is the proper sequence of
these maneuvers?

19. Should extinguishing operations with a
Tower Ladder stream begin on the lower,
middle or upper floor?

20. The names Franklin, Magic Eye or Fox on
a lock cylinder would tell a fireman that
this is what kind of lock?

21. A cylinder in the center of a door of a
commercial occupancy would tell a
firefighter that this is what kind of lock?

22. What type of lock is recessed in the edge
of a door?

23. What is the most common type of door
lock?

24. Which butt is used to connect to a
hydrant?

25. Of all the ways to remove a person from

a stuck elevator, which should be considered first?

26. What kind of survival rate can be expected when a residential building collapses pancake fashion?

27. What is the main hazard at a car fire?

28. What color flame does a large volume of burning alcohol give off?

29. What color flame does a small quantity of alcohol used as an accelerant give off?

30. What are the two leading causes of fires in mobile homes?

31. What is the first thing a firefighter should do at an oil burner emergency?

32. What is an undercut line?

33. What does it mean when a fire "vents itself"?

34. What is the "one lick" method?

35. What does "take up" mean?

36. What does it mean to "knock down a fire"?

37. Who is a line locator?

38. If five out of six ropes were sheared in a traction type elevator, what would be the result?

39. What is the first thing to do when arriving at a gas cylinder leak?

40. At a call for a burning sulphur candle, a firefighter without a mask crawled in to search. Comment on his action.

41. Why is water the most widely used extinguishing agent?

42. Water in what form is effective in dispensing LP-gas vapor?

43. Arriving at a call for a malfunctioning oil burner using #6 oil, what three things must a firefighter shut down?

44. Is it easier to fight a fire in a long narrow warehouse or a large square one?

45. From a firefighting perspective, how many stories high should a warehouse be?

46. Why should the drafting of water from cellars and subways be avoided?

47. What does it mean if a firefighter makes an upward vertical motion of his right arm?

48. When cutting a live wire, why is it important to cut it as close to a stationary object as possible?

49. How should water be applied to a burning liquid to achieve extinguishment by emulsification?

50. Why is it not practical to extinguish a flammable liquid fire in a tank by dilution?

51. How dangerous is it to use a hose stream on a 500 volt wire from a distance of 30 feet?

52. What is the term used when smoke and gases reach the top floor of a building and start to bank down?

53. What is a charged line?

54. Why is it not advisable to use a combination nozzle at an electrical fire?

55. Why should an applicator not be used at an electrical fire?

56. When placing a portable ladder at a window, where should the tip be?

57. How high above a roof should the tip of a portable ladder be placed?

58. What is a firefighter doing when he places one foot on the bottom rung of a ladder and holds both beams?

59. When climbing a portable ladder where do a firefighter's eyes look?

60. When climbing a portable ladder, why aren't tools carried inside the ladder?

61. What is the highest rung on a portable ladder that it is safe to stand on?

62. When descending an aerial ladder with an ambulatory victim, how many rungs below the victim should the firefighter be?

63. Which victim should be brought down an aerial ladder first: a teenager, a child or an old man?

64. Why should the first due ladder company enter a block from the same direction as the first due engine company?

65. Who has the responsibility for the proper placement of an aerial ladder?

66. What should be placed under tormentors or outriggers when they are placed on soft earth?

67. In order to prevent accidental movement

of an aerial ladder, what should be disengaged at extended operations?

68. Which of the following is the safest and easiest way to get to the roof of a fire building: aerial ladder, fire escape or adjoining building?

69. To keep a bulkhead door open, which hinge should be removed?

70. When breaking a skylight for ventilation, why does a firefighter hesitate after breaking the first pane?

71. If while descending a staircase, heavy fire suddenly blows up the staircase, what should a firefighter do?

72. What is the best way to move on a smokey roof?

73. What can premature ventilation do to a fire?

74. At a top floor fire, where should a hole in the roof be cut?

75. What shape should a breach in a brick wall be?

76. What is it called when the chief in charge of a fire estimates the problems and conditions?

77. What is the primary purpose of the first line stretched in a multiple dwelling fire?

78. How can an officer of an advancing hose line monitor heat conditions in a cellar?

79. If potential for a backdraft exists, what type ventilation should be done first?

80. How many strategic factors must be considered when sizing up a fire building?

81. At what stage of a fire should a firefighter be aware of a backdraft possibility?

82. When approaching an LPG fire how should the nozzle be held?

83. In what position should a portable LPG tank exposed to fire be moved?

84. If water spray does not disperse an LPG vapor what strategy should be employed?

85. How should a propane fire be extinguished?

86. True or false: At a hazardous materials incident, quick aggressive action is necessary.

87. Why should sand not be used to cover petroleum spills?

88.   Why should firefighters responding to a hazardous material incident not carry wallets, jewelry, etc.?

Answers

1. Inside the building to protect the interior staircase

2. Stop the flow of gas

3. Ten

4. Cool it, remove the oxygen, remove the fuel, chemical extinguishment

5. A steam explosion

6. Twice

7. The outside electrical connection

8. Negative or ground

9.   Exposures

10.   Cellars and storage areas

11.   Drop to the floor and lie flat

12.   Shake it to determine weakness

13.   Three

14.   Bathroom

15.   Bearing wall

16.   A wall in a five story building

17.   35'

18.   Raise, rotate, extend

19.   Lower

20.   Police vertical bar lock

21.   Fox police double bar lock

22.   Mortise lock

23.   Rim lock

24.   Female

25. Through the car and hoistway doors

26. A high survival rate

27. The gas tank

28. Orange

29. Blue

30. Heating and electrical distribution

31. Shut down the oil burner at the remote control switch

32. A fire line cut below fire on a slope

33. Fire has penetrated to the outside of a building

34. A method of constructing a fire line in a forest fire

35. Leave the fire scene

36. To reduce the flames and keep the fire from spreading (also called darken down)

37. The person who decides where a fire line will be started in a forest fire

38. The remaining rope would hold the car

39.   Determine the type of gas

40.   Sulphur gas is heavier than air and there will be higher concentrations near the floor. It is not wise to enter without a mask when sulphur gas is present.

41.   Because of its high heat absorption capability and its availability

42.   Spray or fog

43.   Remote electrical switch, oil ow valve, pre-heater valve

44.   Long narrow

45.   One

46.   Foreign matter may damage pumps on apparatus

47.   Start water

48.   To prevent whipping

49.   Strong, coarse, straight stream

50.   Large amount of water needed, danger of overflow

51.   Little danger involved

52. Mushrooming

53. A hose line with water in it

54. Danger of accidentally using a straight stream instead of fog

55. Danger of contact with the electrical equipment

56. Level with the window sill

57. At least two feet

58. Butting the ladder

59. Up or forward

60. If dropped they will hit the butt man

61. Third rung from the top

62. One

63. Teenager (he or she will be able to climb fastest)

64. So as not to be blocked from the front of the fire building by the engine company

65. The officer in charge of the ladder company

66. Planking

67. The power take off

68. Adjoining building

69. Upper one

70. To allow firefighters below to get out of the way

71. Drop to the floor and roll to the wall

72. In a crouched or kneeling position

73. Increase its intensity

74. Directly over the fire

75. Triangular with the vertex angle upward

76. Size-up

77. To safeguard the stairway

78. Extend a bare hand overhead

79. Roof or vertical

80. Thirteen

81. Third or smoldering stage

82. Low and aimed upward

83. In an upright position

84. Use heavy streams of water from a safe distance

85. By shutting off the flow of escaping propane

86. False

87. To allow for recovery

88. They may become contaminated and have to be impounded

Fire Science

On the surface, fighting a fire seems like a fairly simple thing to do. Put water on it and the fire will go out. If that was all there is to it, firefighting would indeed be an easy profession. Actually, it is a complex, technical field requiring many hours of study.

It is not necessary to have a complex knowledge of fire science in order to fight a fire. In fact, having knowledge of the basics is often enough to get the job done. But somebody at the fire scene should know what to do when the unusual occurs. This is where the fire science courses become very useful.

With the increasing use of chemicals and plastics, it is imperative that firefighters know exactly what they are dealing with at the fire scene. Many of these plastics are carcinogenic in their normal state. Imagine the health hazards that they pose when they are burning. Our main asset at these types of fires is the knowledge that we have obtained through study. The practical experience of firefighting is very important but so also is the knowledge of fire science.

A few years ago I decided to study for promotion to the rank of Lieutenant. I enrolled in a course given by a private school. This school had been preparing firefighters for promotion for many years and had a high success rate. They focused not so much on fire science as they did on management techniques, evolutions and fire department bulletins and circulars.

One of the younger men in the company also decided to study for promotion but instead of enrolling in the same school, he pursued a degree in fire science at John Jay College. I told him that he was making a mistake by taking such courses because the facts that he learned would not be asked on the promotional exam.

Well, I was right in the sense that most of what he studied was not asked on the test, but he managed to write a much higher mark than me and today he not only has a degree but will soon be promoted to the rank of Battalion Chief. So perhaps, I wasn't right after all.

These questions are the most difficult in this book do don't be discouraged if you have trouble with them. If you find them easy, you should be named the Chief of Department.

Fire Science

1. In what three ways does heat leave a fire?

2. What kind of solid is grease?

3. What is the approximate ignition temperature of wood?

4. What are the three main chemical ingredients of most solids?

5. When can wood ignite spontaneously?

6. What is the approximate temperature of a match flame?

7. What is the approximate temperature of a lighted cigarette?

8. In what two ways can a liquid become a gas?

9. A Class 1 liquid has a flash point below what temperature?

10. Combustible liquids have a flash point above what temperature?

11. If liquid hydrocarbons were burning what would be the visual result?

12. If alcohol was burning, what would be the visual result?

13. What gases will not support combustion?

14. What is a cryogenic gas?

15. What is the difference between a nitrate and a nitrite?

16. What stage of fire is the incipient stage?

17. In a sprinkler system, what valve allows sections of pipe to drain?

18. Why were carbon fibers originally produced?

19. At what temperature does liquid propane boil?

20. How many times does water expand when changing from a liquid to a vapor?

21. What is another name for a glass or frangible bulb type of sprinkler head?

22. What organization in 1953 improved the deflectors in sprinkler heads?

23. What is an open head system?

24. What is a vertical water pipe riser in a building which supplies fire hoses?

25. Is #2 heating oil considered a light or heavy oil?

26. What does a pressuretrol on an oil burner do?

27. What does an aquastat on an oil burner do?

28. What two types of automatic wet sprinkler systems are there?

29. What two gases are present in ammonia?

30. What is ethylene gas mainly used for?

31. What is the lightest known gas?

32. What kind of gas is found in most city gas mains?

33. What is the principle ingredient of natural gas?

34. When referring to a gas, what do the letters LPG stand for?

35. What should be the minimum width of a main aisle in a warehouse?

36. What kind of coal is prone to spontaneous heating?

37. Are most flammable liquids lighter or heavier than air?

38. What is the most commonly used flammable liquid?

39. What is JP-4 fuel usually used for?

40. What is the life expectancy of an underground gasoline tank?

41. At what temperature is air liquefied?

42. What is a rapid instantaneous combustion of flammable gases which explodes called?

43. Catalytic converters have been known to start what type of fire?

44. What gas does carbon tetrachloride form when it comes in contact with a heated metal?

45. How does hydrogen sulphide cause death?

46. On what type of solid fire is Halon 1301 not effective?

47. What type of radiation is long range and very penetrating?

48. What type of radiation is an external hazard but can be stopped by material such as aluminum?

49. What type radiation is only an internal hazard?

50. Which cells of the body are most harmed by radiation?

51. Which cells of the body are least affected by radiation?

52. What would be considered an acute exposure of radiation?

53. What are measurements of radiation called?

54. What are the two most common types of radiation hazard?

55. For whom was the roentgen named?

56. What does a transformer do?

57.　What is transformer oil called?

58.　If a person was exposed to 75 roentgens of radiation, how would his health be affected?

59.　Which type alarm device will operate first in a fire: heat sensitive, smoke activated, or flame enegry?

60.　What is pyrolysis?

61.　What are the two general categories of heat activated alarms?

62.　What is the advantage of a snap action disc thermostat over a bimetallic strip thermostat?

63.　What is the greatest disadvantage of a rate of rise fire alarm compared to a fixed temperature device?

64.　What is the main reason for using a timing device in setting a fire?

65.　What four groups of people turn in most false alarms?

66.　What three high explosives are most frequently encountered by firefighters?

67.　Wht do the letters TNT stand for?

68. What is the easiest crime to commit and the hardest to detect?

69. What must be eliminated before a fire can be classified incendiary?

70. What odor does burning phosphorous give off?

71. What is the first thing to determine when investigating the cause of a fire?

72. When investigating a fire, who should be questioned first?

73. In arson investigation, what does the term "plant" mean?

74. What is the most common type of safety device on a compressed gas cylinder?

75. What is static pressure of water?

76. In a hose line what is greater: static pressure, flow pressure, or residual pressure?

77. What temperature is absolute zero?

78. When gas molecules become sluggish and tend to adhere to one another, what does the gas become?

79. If gas molecules stick together, what does the gas become?

80. What is the process called whereby a gas spreads out and uniformly fills a space?

81. What is the atmospheric pressure at sea level?

82. What law states that if the temperature remains constant, the volume of a confined gas varies inversely as the pressure applied to it?

83. What law states that the volume of a gas at constant temperature is directly proportional to the absolute temperature?

84. What type gas would be in a cylinder with a green label on it?

85. What type of gas would be in a cylinder with a red label on it?

86. What two safety devices do all flammable gas cylinders have?

87. What safety device is on a nonflammable gas cylinder?

88. Where do LP gases come from?

89. What is petroleum mainly composed of?

90. What is the correct color of an oil burner flame?

91. What are three types of forest fires?

92. If barium peroxide is burned, what common gas is released?

93. What is the color of sulfur?

94. What is the only kind of coal that can not ignite spontaneously?

95. What are the two types of phosphorus?

96. What is the color of fluorine?

97. What is the color of chlorine gas?

98. What are the three most common causes of an explosion?

99. What primarily determines the destructive effect of a dust explosion?

100. If burning magnesium is hit with water, then carbon dioxide, then nitrogen, what will it do?

101. What is the melting point of steel?

102. What do the letters PVC stand for?

Answers

1. Radiation, conduction, convection

2. Amorphous

3. 385-400 degrees

4. Carbon, hydrogen, oxygen

5. When it has changed to charcoal

6. 2,000 degrees or higher

7. 525-1300 degrees

8. Its temperature is increased or pressure decreased

9.   100 degrees F.

10.   139 degrees F.

11.   Orange flame, dense black smoke

12.   Blue flame, little smoke

13.   Inert gases

14.   A gas that has been liquefied and stored at a low temperature

15.   Nitrites have one less oxygen atom

16.   Initial or first

17.   Ball drip valve

18.   As an incandescent lamp filament

19.   -45 degrees F.

20.   1700

21.   Quartzoid

22.   Factory Mutual

23.   A sprinkler system whose heads do not have fusible links

24.   Standpipe

25.   Light

26.   Gauges high and low pressures and activates or deactivates the system

27.   Monitors the water temperature in a hot water system and activates or deactivates the system

28.   Variable pressure and constant pressure

29.   Nitrogen and hydrogen

30.   Ripening fruits

31.   Hydrogen

32.   Natural gas

33.   Methane

34.   Liquefied petroleum gas

35.   8 feet

36.   Bituminous coal

37.   Heavier

38.   Gasoline

39.   Jet aircraft

Fire Science—Answers

40.   20 years

41.   -317 degrees F.

42.   Backdraft, blow-back, or smoke or hot air explosion

43.   Brush and leaves

44.   Phosgene

45.   Paralyzes the nervous and respiratory system

46.   Those solids which provide their own oxygen

47.   Gamma

48.   Beta

49.   Alpha

50.   White blood cells of the spleen and lymph nodes

51.   Nerve cells

52.   25 roentgens or more at one time

53.   Dosimetry

54.   X-rays and gamma rays

55.   Dr. Wilhelm Roentgen

56.   Steps electrical current up or down

57.   Transil oil

58.   No radiation sickness, but slight temporary changes in the blood

59.   Smoke activated

60.   Chemical decomposition of matter by heat

61.   Fixed temperature and rate of rise

62.   Fewer false alarms

63.   If the temperature rise is gradual, the alarm will not be triggered

64.   To establish an alibi

65.   Children, drunks, mental defectives, firemen

66.   Dynamite, TNT, nitroglycerine

67.   Trinitrotoluene or trinitrotoluol

68.   Arson

69.   All accidental causes

70. Garlic

71. Where did the fire start?

72. The discoverer of the fire

73. Preparation for the unlawful setting of fire

74. Fusible plug

75. Pressure of water when it is not moving

76. Static pressure

77. -459 degrees F.

78. Liquid

79. Solid

80. Diffusion

81. 14.7 psia

82. Boyle's Law

83. Charles' Law

84. Nonflammable gas

85. Flammable gas

86. Frangible disk and fusible plug

87.   Frangible disk

88.   Oil and natural gas wells and oil refinery processes

89.   Carbon and hydrogen

90.   Orange with small red tips

91.   Ground, surface and crown

92.   Oxygen

93.   Yellow

94.   High grade anthracite

95.   White or yellow and red

96.   Greenish yellow

97.   Greenish yellow

98.   Shock, heat, pressure

99.   The rate of pressure rise

100.   Continue to burn

101.   2,600 degrees F.

102.   Polyvinyl chloride

Odds and Ends

There must be a place for Smokey the Bear and Sparky the firedog, and this is it. These are the "off the wall" questions that you have been waiting for. Everyone from John Wayne to Saint Florian is in this chapter. These are the most trivial of all the trivia questions in the book.

In researching and writing this book, whenever I happened upon an item that did not fit into any of the previous categories, I put that item into this section. I found these questions to be the most interesting and fun. See if you agree.

Odds and Ends

1. What is a relief pitcher in baseball called?

2. What does "stand fast" mean?

3. What is a four bagger?

4. What is the slang term for an arsonist?

5. What is a recall?

6. What is a rekindle?

7. What is a route card?

8. Who is the patron saint of firefighters?

9. What does moonlighting mean?

10. What former publication of the NFPA was published four times a year?

11. What does NFPA stand for?

12. What U.S. President pardoned the captain of the General Slocum?

13. What is a hotshot crew?

14. What is the firefighter who turns out the company called?

15. Who is an incendiary?

16. In firefighting, what does the term "head" refer to?

17. In firefighting, what two things does the term "hook up" mean?

18. What is a fire fan?

19. What is a friendly fire?

20. How many U.S. gallons are in an Imperial gallon?

21. In what year was the Fire Protection Handbook first published?

22. Who were the three authors of the first seven editions of the Fire Protection Handbook?

23. What is the term used when a firefighter is sent to another company for a tour?

24. Who is the editor of *Firehouse* Magazine?

25. What breed of dog is known as a firedog?

26. What is a pyrotechnist?

27. What is a pyromaniac?

28. Why do people burn sulphur candles in their apartments?

29. Why is it usually a woman who sets a fire in a closet?

30. What is pyrophobia?

31. What is pyrosis?

32. What is a pyrostat?

33. What army adopted the Maltese Cross as its badge?

34. What group of men first used the Maltese Cross as its symbol?

35. What does the color red on a badge signify?

36. What does an eagle on a badge symbolize?

37. What part of the body is a sentry dog trained to attack?

38. A firefighter assigned to a fireboat would know that the starboard side is which side?

39. A firefighter assigned to a fireboat would know that the port side is which side?

40. Who rebuilt London after the Great Fire?

41. Where is the American Museum of Firefighting located?

42. How many times a year does Fire Technology publish?

43. Where is the main office and test station of Underwriters' Laboratories?

44. What is the most dangerous profession in the U.S.?

45. What is the name of the club whose members collect insurance marks?

46. What law was passed after the Triangle Shirtwaist Fire?

47. Why did firefighters prefer dalmations over other breeds of dogs?

48. What famous comedian was due to perform the night of the Iroquois theater fire?

49. From what country did dalmations come from?

50. Which god did the ancient Greeks believe gave them the gift of fire?

51. When did man first use fire for heat?

52. Who invented the friction match?

53. Why are dalmations plagued by physical problems such as deafness?

54. What is Operation Edith?

55. What is the official name of the Maltese Cross?

56. What is the Maltese Cross named for?

57. What will the Phoenix rise out of?

58. What is the first line of the Fireman's Prayer?

59. What does Smokey the Bear say?

60. Where was St. Florian born?

61. When is Fire Service Recognition Day observed?

62. What day is Firefighters Memorial Sunday?

63. What week is Fire Prevention Week?

64. Who proclaimed the first Fire Prevention Day?

65. Who proclaimed the first Fire Prevention Week?

66. What does a fire helmet thrown from a window mean?

67. Who made the famous print "Life of a Fireman"?

68. What type of wood does not have to be seasoned before being burned in a fireplace?

69. Dr. Martin Luther King's death precipitated riots and arson fires. What is the date of his death?

70. What city was known as the "Fire Engine Capital of the World"?

71. What does it mean to be an acting Lieutenant?

72. What is a bushwacker in the fire service?

73. What is a chowder fireman?

74. In what city is the International Fire Buff Associates located?

75. When was Sparky the firedog born?

76. When was Nomex(R) invented?

77. With what did Red Adair extinguish the greatest gas fire in history?

78. How much money was Red Adair paid to extinguish the greatest gas fi e in history?

79. In what city were 800 peo e killed by a fireworks display?

80. In what country was the earliest evidence of use of fire by man found?

81. In what country was the brightest fire photographed?

82. Who extinguished 8393 flaming torches in his mouth in two hours?

83. In what year was the International Association of Firefighters founded?

84. What is the record for the most people killed by a single lightning bolt?

85. What is the name of the largest firework ever made?

86. Who fiddled while Rome burned?

87. In the movie Turk 182, who played Fireman Terry "Turk" Lynch?

88. In the movie Turk 182, what did the number 182 signify?

89. In Penny Lane there is a fireman with an hourglass. What is in his pocket?

90. How many fires have there been in the White House since the War of 1812?

91. What President was throwing a Christmas Eve party when the White House went on fire?

92. What is the difference between flammable and inflammable?

93. What is the French word for the firefighter?

94. Where is the locking device of a subway turnstyle usually located?

95. Why is Fire Prevention Week always the Sunday through Saturday in which October 9 falls?

96. What is the best type of wood to use in a fireplace?

97. When a battery is being charged, what two gases are present at the top of the battery?

98. What country produces over 50% of the world's carbon monoxide?

99. What is the weight of a subway car?

100. How do you signal a train to stop with a flashlight?

101. If a train engineer saw a green light on the ties between the rails, what would he do?

102. In what five cities are insurance salvage companies still operating?

103. In a residential area what is the maximum distance a person should have to travel to a fire alarm box?

104. What type fire alarm system notifies only the occupants of a premise?

105. A WFAS alarm is received. What does WFAS stand for?

106. What government agency regulates ionization detectors?

107. How many volunteer fire companies are there in N.Y.C.?

108. What is an evolution?

109. What are fire lines?

110. What powers catenary powered trains?

111. What are the shipping papers called which list what is carried on a truck?

112. What are the shipping papers called which list what is carried on a train?

113. Who carries the shipping papers on a railroad train?

114. What are the shipping papers called which list what is carried on an airplane?

115. Who carries the shipping papers on an airplane?

116. What are the shipping papers called which list what is carried on a ship?

117. Who carries the shipping papers on a ship?

118. Who was the Roman god of fire?

119. In what country were safety matches invented?

120. In what religion does a widow throw herself on her husband's cremation fire?

121. What city hosts the oldest firemen's parade?

122. What will the weather be like on the day of the Port Jervis N.Y. Firemen's Parade?

123. How did Fire Island N.Y. get its name?

124. What is the common name for a lightning bug?

125. What do a chimney, mantel and hearth comprise?

126. What Greek philosopher believed that fire was the primal element?

127. What saint, known as the Maid of Orleans, was burned at the stake?

128. What was the most widespread means of starting a fire among primitive people?

129. A Dakota Indian legend says what animal gave fire to man?

130. What kind of fire is made in the Roman Catholic Church during Passion week?

131. In what building was the Triangle Shirtwaist Co. located?

132. How many airship disasters were there between 1880 and 1938?

133. Where is the oldest fire station in America located?

134. Name the three now deceased stars of "Hellfighters".

135. What was fire first used for?

136. What is St. Elmo's fire?

137. In what firehouse was "Ghostbusters" filmed?

138. Which state was bombed with incendiaries by the Japanese during World War II?

Answers

1. Fireman

2. Standby; be available to work at a fire scene

3. Four alarm fire

4. Torch

5. Calling off duty firefighters to work in an emergency

6. Fire thought to be extinguished, reignites

7. Index card with directions to a certain location

Odds and Ends—Answers

8.   St. Florian

9.   Firefighters working another job

10.   The Quarterly

11.   National Fire Protection Association

12.   William Howard Taft

13.   Forest firefighters

14.   Housewatchman

15.   Person who deliberately sets a fire

16.   Back pressure; pressure due to elevation of water

17.   Connect a pumper to a hydrant; formal disciplinary action against a firefighter

18.   Buff

19.   Fire lit intentionally for a good purpose

20.   1.201

21.   1896

22.   Crosby, Fiske, Forster

23.   Detail

24. Dennis Smith

25. Dalmation

26. Fireworks expert

27. Fire-bug; one who has a persistent impulse to burn things

28. To kill bugs, particularly cockroaches

29. Desire for a new wardrobe

30. Morbid fear of fire

31. Heartburn

32. Device which gives a warning when exposed to fire

33. Union Army of the Civil War

34. Christian Knights of the Crusades

35. Courage and valor

36. Authority

37. Hand holding a weapon

38. Right side looking forward

39. Left side looking forward

40. Christopher Wren

41. Hudson, N.Y.

42. Four

43. Chicago

44. Firefighting

45. Fire Mark Circle of the Americas

46. New York State Labor Law

47. They are a coach dog and get along well with horses.

48. Eddie Foy

49. Yugoslavia (Dalmatia Coast)

50. Prometheus

51. At least 500,000 years ago

52. An Englishman, John Walker

53. They were inbred by firefighters, particularly in New York City

54. Exit drills in the home

55. The Cross Pattee-Nowy

56. Island of Malta

57. Its own ashes

58. "When I am called to duty, God, wherever flames may rage, give me strength to save some life, whatever be its age"

59. "Only you can prevent forest fires"

60. Austria

61. Second Saturday in May

62. First Sunday of October

63. The week Sunday to Saturday which includes October 9

64. President Woodrow Wilson (October 9, 1920)

65. President Warren G. Harding (1922)

66. Firefighter in distress

67. Currier & Ives

68. Ash

69. April 4, 1968

70. Seneca Falls, N.Y.

71. Firefighter of next lower rank is detailed to perform the duties of Lieutenant

72. Firefighter in a suburban area fighting brush fires

73. Inactive or honorary firemen who show up at all the parties

74. Baltimore

75. 1951

76. 1967

77. Dynamite

78. One million dollars

79. Paris

80. Kenya

81. Czechoslovakia

82. Reg Morris of Great Britain

83. 1918

84. 21 (Rhodesia)

85. Fat Man II

86. Nero

184

Odds and Ends—Answers

87. Robert Urich

88. Terry Lynch's badge number

89. A portrait of the Queen

90. Five

91. Herbert Hoover

92. No difference

93. Pompier

94. In the floor base under the centerpost

95. October 9 is the anniversary of the Great Chicago Fire.

96. Hickory

97. Hydrogen and oxygen

98. U.S.A.

99. 40 tons

100. Horizontal movement

101. Stop the train

102. N.Y.C., London, Liverpool, Glasgow, Bombay

103. Two blocks or 800 feet

104. Local alarm system

105. Wells Fargo Alarm System

106. Atomic Energy Commission

107. Ten

108. Predetermined firefighting maneuver

109. Boundary set up by police to keep civilians away from the fire scene

110. Overhead electrical lines

111. Bill of lading

112. Waybill

113. Conductor

114. Air bill

115. Pilot

116. Cargo manifest

117. Master or first mate

118. Vulcan

119. Sweden

120. Hinduism

121. Port Jervis, New York

122. Clear—it has only rained once on the parade in its 135 year history

123. Early inhabitants lit fires to lure ships to destruction.

124. Firefly

125. Fireplace

126. Heraclitus

127. Joan of Arc

128. Friction

129. Panther

130. New fire

131. Asch Building (Green St. & Washington Pl., NYC)

132. Two (British R101 and Hindenburg)

133. Mount Holly, New Jersey

134. John Wayne, Jim Hutton, Jay C. Flippen

135. Warmth

136. Flamelike appearance seen on ships masts during storms

137. N.Y.C.'s Ladder Co. 8

138. Oregon